I hope Gareth Jones
enjoys reading this book

AS TIME GOES BY

Robert Jones

MINERVA PRESS

LONDON

MIAMI   RIO DE JANEIRO   DELHI

AS TIME GOES BY
Copyright © Robert Jones 2001

All Rights Reserved

ISBN 0 75411 471 6

First Published 2001 by
MINERVA PRESS
315–317 Regent Street
London W1R 7YB

Printed in Great Britain for Minerva Press

AS TIME GOES BY

Contents

Clocks

The Passing of Time

The Lords

Simple Pleasures

Clocks

Another Watch Tale

Another tale we heard today,
About the master's watch,
That marmalade cat from 'Meadowlea',
Had seen him scratch his thatch.

She told this to the tomcat,
Who related it to us,
The master doing his walkabout,
Had focused on his watch.

Having raised his arm, and moved his cuff,
He viewed an empty wrist,
This put him in a right old huff,
He was not expecting this.

Had hoped to see his quartz watch,
The calendar, the time,
But now must carry out a search,
Back to the 'Dog and Gun'.

The spaniel did not see the point,
Of traipsing back to town,
'An excuse to have another pint',
But master called the tune.

The dog voiced no opinion,
But displayed a bleary eye,
The master in high dudgeon,
Muttered nasty words to Roy.

He wished he had a 'sniffer' hound,
One trained to find a watch,
Thus avoiding all the ups and downs
And scratching of his thatch.

We are told he never found the quartz,
Despite extensive hunting,
And now he's back with his old watch,
An ageing gold half hunter.

The Turnip Watch

The master speaking to old friends,
Recalled the good old days,
All the familiar sights and sounds,
Oh how those memories stay.

You cannot beat the good, loud tick,
Of my old turnip watch,
A solid job, has had some stick,
It licks the hide off quartz.

The spaniel sneered, remembering how,
When master lost the quartz,
The fuss there was, and all the row,
The searching in the gorse.

He did not tell these friends of his,
About this episode,
He harped about his turnip watch,
No mention of those words.

The nasty words he said to Roy,
As though he was to blame,
He told them though, that when a boy,
Dad had a smart Great Dane.

A really clever dog, this Nip,
Could tell the time and count,
Would look at master's turnip,
Tell him, 'Time for walkabout'.

This made the spaniel quite upset,
Poor Roy was very glum,
He wished he was the tomcat,
Who did little else but yawn.

What noise was that the tomcat asked,
As something heavy dropped,
Must be the master's turnip watch,
And I think that it has stopped.

The turnip watch was on the floor,
The master was aghast,
The spaniel seeking cover,
Had overturned a vase.

The mistress now, was on the scene,
Smashed vases lure the dames,
How came you, Jack, to break that Ming?
The dog was not to blame.

The spaniel was astonished,
By this sudden change of luck,
He thought he'd be admonished,
And possibly be locked up.

What the Wall Clock Saw

The mistress tore a strip off Jack,
That day he broke her favourite vase,
She put him truly on the rack,
Told him to mend his foolish ways.

He blamed the spaniel first of all,
She did not believe his story,
I saw what happened from this wall,
I agreed with master's theory.

He was reaching for his glass of mixed,
And rambling on about his watch,
When suddenly the watch escaped,
Then followed this most frightening crash.

The vicar, who was sipping stout,
Saw master's watch dive to the floor,
He leapt three foot, despite his gout,
And promptly hid behind the door.

You could not blame the parish priest,
Nor could you fault poor Jack,
The blame lay with this clumsy beast,
Who gave her vase a whack.

This done, he broke from cover,
Then hastened to the door,
He said, 'The party's over,
I think I'll move next door.'

Lancashire Watches

Long years ago, the Higginsons
A watch made, at St Helens,
'Twas in the year one six nine seven,
It was indeed a good one.

This watch was such a special piece,
That all the county sought one.
According to the book by Weiss,
They made some more in Wigan.

The master mentioned Liverpool.
He also listed names and towns,
Mathew Gleave and Thomas Poole,
West Houton, Ormskirk, Warrington.

The county now made all the parts
That go into a watch,
Brass wheels and other works of art.
The master said the brass was Dutch.

The vicar and old Blore were there.
These two did not agree.
Blore thought the springs were made of hair,
The pinions made of steel.

The Bracket Clock

Tom Tompion, many years ago,
Produced his famous bracket clock,
'Twas in the year one six one four
Tom gave the clock world quite a shock.

He put his country to the fore,
As makers of the best of clocks.
With pendulums swinging to and fro,
His beat the Frenchmen, Swiss and Boche.

The watches too caused quite a stir,
Hair sprung, were neat and pretty.
They kept good time, were better far,
Than those made in the smithy.

The blacksmith now made shoes and ploughs,
With heavy blows grew mighty arms,
Made metal gates, and pumped bellows,
Forged implements for all the farms.

The Anniversary Clock

'Twas told to me the other day,
A tale about a clock,
Anniversary type they say,
Had once been put in hock.

Degrading places, pawn shops,
Take you down a peg or two.
This clock was once a bishop's,
He bought it when 'twas new.

Sad to say, disaster struck him,
For he was on the rocks.
His creditors were pressing,
So he chose to pawn the clock.

Worse still, did not redeem it,
Left it languishing in pawn,
Until one day old Hacket,
The eccentric from the 'Lawns',

Took a fancy to the glass domed freak,
And bought it for one pound,
Thinking he had scored off Peek,
And thought the clock was sound.

Peek's tomcat knew a thing or two,
For tomcats get about.
He said 'That clock will never go,
That is why the bishop slung it out.'

The Quartz Watch

The master going to catch a bus,
Was looking at his watch.
He did not see the tabby puss,
Tripped up and hurt his thatch.

The master and the tomcat,
Were both a trifle piqued.
The master dislikes plasters.
Tomcats hate being kicked.

Silicon dioxide, I blame for all the fuss.
'Tis packed inside his wretched watch,
I heard that from the puss.
The tomcat also hinted that it is called a pulsar quartz.

Digital and faceless, that's no way to show the time.
The system is abhorrent to self-respecting clocks,
A battery, crystal combine,
Oh how the wall clock mocks.

The Cuckoo Clock

I seemed to startle dog and cat,
When first I came, a job to do.
I've never seen them move as fast,
As when they heard me shout, 'Cuckoo'.

The other clocks all think they're it,
Some are old, some boast, 'I'm new,'
I think they try to imitate.
Excuse me, while I shout, 'Cuckoo.'

That wall clock chimes each half and full.
The master bought him years ago.
As clocks go, he's rather dull.
Excuse me, while I shout, 'Cuckoo.'

It looks as though he has dry rot.
Down in the mouth, I think, don't you?
There is no hope for clocks with rot.
Excuse me, while I shout, 'Cuckoo.'

That tomcat peeping from the hall,
Has crouched like that since half past two,
But I feel safe high on the wall.
Excuse me, while I shout, 'Cuckoo.'

The other clocks all tick, some chime.
My behaviour may seem odd to you,
With interruptions all the time.
I'm duty bound to shout, 'Cuckoo.'

Round Head Clock

The master has another clock,
A 'pine' we heard him say,
Another bargain bought from hock,
It came here, yesterday.

We heard the vicar tell old Blore,
A round head not a pine.
Blore replied, 'I'm not so sure,
There's one like that at "Dog and Gun".'

I hope they leave me on this wall,
Well out of reach of tomcats.
The spaniel heard them mention hall.
There is little there to look at.

The hall is not the place for clocks,
Unless they are 'long case',
I tick, shout cuckoo, and I tock,
But cannot match grandfather's face.

The Passing of Time

In the Beginning

As old as time, deep darkness stood,
Until heaven's lamps were lit by God.
Vast waters which had clothed the earth,
Began to live, were given breath.
These waters now could see by day,
The moon ruled o'er their tidal way.
New lands appeared, whose shores they lap,
From Europe to the polar cap.
Southward they reach to penguin land,
Past Africa's palm-fringed golden sand.

This was the dawn of history,
With oceans, cloaked in mystery,
Who gave a home to gull and whale,
Before they saw a boat or sail.
They felt the first breath of the wind.
Sometimes it purred, at times it stormed,
Pure, these waters, free from filth,
For man had yet to strive for wealth.
Small too was Adam's family tree,
And strange to him the fretful sea.

Who launched Phoebus on his way,
To lustre bright the face of day,
Designed the heavens with stars and moon,
The seasons dancing each in turn,
Explosive lightning's fiery hand,
To terrorise and burn the land.
The humble poet wracks his brain,
These mighty mysteries to explain.
The scientist speaks foolish word.
He tries to prove there is no God.

Heaven

Friend, know you not that heaven is for all,
Its vestibule is cluttered with tiaras, crowns and chains,
A club with membership embracing all,
Ex-paupers sit with former kings and queens.

Here each a scholar is, and understands
The common language, sings the songs.
Here all rejoice and wish not for a magic wand.
They bask in freedom from ideas and wrongs.

New understanding makes each one aware of all techniques,
Now all can see, can hear, can speak and read,
Unfolded now the mystery of magic tricks,
Each member plucks a harp in praise of God.

Time here is measured in one way,
Gone are the days, the weeks and years,
One unit serves, a long heyday,
Of endless pleasure free of pain and tears.

Handel's Flute

Played in heaven, so I believe,
The sweeter flute of Handel's time.
A simple tube of wood, no key,
Our modern flute is less sublime.

Gone from the earth has Handel's flute.
He took it with him into Heaven.
Sad that we have to substitute
With metal, gadget-strewn, machine.

Perhaps, great Handel there presides,
O'er oratorio, flute and choir.
Here metal, key-bound flute abides,
On earth we listen to, 'Messiah'.

The Obscure

One of Egypt's pyramids is called 'The Great Pyramid',
Perhaps in memory of the slaves who built them.

Cyrus conquered Babylon.
Alone? Clever man.

Dionysus gave Ariadne a wedding garland.
Who grew the flowers?

Rome was not built in a day.
Who built it and who supplied the material?

Sir Francis Drake defeated the Spanish Armada.
How many shots did he fire?

On September 13 1759, General Wolfe fell at Quebec.
The only casualty?

General Elliot was raised to the peerage, for his brilliant defence,
Of Gibraltar, against the Spanish and French.
What happened to his batman?

Of Passchendaele, history records,
The men were obscurely great.
The command obstinate and commonplace,
Never did so vast a mass of brave men,
Drag itself to death and wounds,
Through a gulf so squalid.
Who was the great man who arranged this event?

Sir Bernard Montgomery was a great general.
I knew many of his infantry.
They were all great men.
Let us who enjoy peace, salute them.

History is littered with great men.
They are usually called admiral, general or archbishop.

Victory

They faced the foe with quiet pride.
Fierce storms of battle they did ride,
Then many thousands of them died.

The infantry, from muddy earth,
Crept out at dawn to prove their worth,
Through heavy fire they went forth.

The thunder of the guns did drown,
The marching foot, the mortal wound,
Mad cannonade the only sound.

Brave men of Kent, of Yorks and Lancs,
They hasten on to swell the ranks,
Of we who charge his guns and tanks.

To Caen, Falaise, across the Rhine,
The foe retreats, to us the prize
Of peace, crutches, wounds and blinded eyes.

Bersham Colliery

Far back in time before my days,
One hundred years or more they say,
Tough miners cut through rock and clay,
Far down a shaft they made their way.

They faced grave danger in this well,
Oft angry earth bruised weary bones,
Dug deep and made their way to hell,
Their lot became night's lasting frown.

Condemned to hell, the slave girl and boy,
Cast into pit, exploited and bruised,
Denied the good life our children enjoy,
Toiling in darkness, ill-fed and ill-used.

Long hours they toiled, away from the sun,
Breathed dangerous gas and poisonous dust,
Whilst the owners enjoyed a life full of fun,
The slaves suffered injury, hunger and thirst.

Men, women, and children endured the curse,
Of the charter man who ruled by the rod,
Silicosis and asthma, often much worse,
One ton of coal, cost a pint of their blood.

God's mills grind slowly, they see justice done,
This pit is to close, this section of hell,
The victims escape to fresh air and sun,
Now comes their turn to play and look well.

Red Tiles

Red tiles, the caps of cosy cotts',
Of schoolroom, manse and village hall,
The never failing shield, that shuts
Out cold, heat, wintry snow and hail.

Red tiles, which please the traveller's eye
As, homeward bound, he views his home,
Rooks dance and ride a rainy sky,
They play a game before the storm.

Red tiles, unmoved by rain or blast,
Glance at the restless weather vane,
The cock points south, and then points west,
His work today is quite a strain.

Red tiles, that many stories tell,
Spectators of the rural year,
Recall the building of the mill,
Whose flailing arms cleave restless air.

Red tiles, that on the schoolroom sit,
Wonder why their neighbour 'neath the elms,
Wears sombre coat and slate-grey hat,
On Sunday speaks to God, and sings sweet psalms.

Red tiles, atop the 'Dog and Gun',
Can see afar the beach, the seas,
Tell tales of sailing ships, that run
The Channel's course before the breeze.

Red tiles, that shelter cottages and farm,
Look at the tranquil rush-fringed pool,
The children of the hamlet playing games,
As tomorrow's world makes its way to school.

Red tiles, that crown the distant palace towers,
See strutting peacocks on the palace walks,
A tapestry of lawns and myriad flowers,
The timeless calendar of God at work.

Yesterday

Yesterday, the day you remember best,
The day when everything was better,
The day you talk about most.

Yesterday, was a brilliant summer day.
In pretty frocks and sandals,
Ladies sauntered fragrant, floral ways.

Yesterday, was a lovely day.
We who today will compare.
At sunset what will we say?

Yesterday, Caesar was mighty, all powerful.
Today he is no more,
But sunsets are still wonderful.

Yesterday, dictators ruled with iron hand,
Made days and nights fearsome and fiery,
Strangely that is nostalgic and grand.

Yesterday, the world was full of gentlemen and schoolboys,
And there was one pretty girl, my teacher.
Men worked and made very little noise.

Silence Is

The departure of daylight,
The stealthy approach of twilight,
The descent of the cloak called night.

When the wind tires and sleeps,
A calm untroubled deep,
A smoothly gliding ship.

Words wisely not said,
Anger locked in your head,
Peace reigning instead.

The glow of her presence last night,
When hands touched in delight,
Our unspoken, 'Goodnight'.

Looks exchanged across a room,
With her in mind, meditation,
And continuous imagination.

Before the night has flown,
The waiting for dawn,
The unplucked lyre of morn.

Evening

With murmuring of distant tides on lonely shores,
At the day's ending, with now completed chores,
The evening greets night's lamplight and closing doors.

A cosy hearth its loved ones now enfolds.
Strange forms and pictures, the burning coals unfold.
The pleasures here, more precious are than gold.

Come has the time for pleasure and repose,
Reviving food and warmth, cosy slippers replacing shoes,
A pre-bedtime romp, the playground brought indoors.

For a while, play, then Mum takes him upstairs,
Now, hush, hush, as John says his prayers,
A goodnight kiss, now sleep, clutching his teddy bear.

Night

The purple evening, pale, departs,
And music of the day is stilled.
Romantic airs affect the hearts,
Of shepherd swains and pretty girls.

The twilight softly cloaks their world.
It walks them gently into night.
Though airs are now a little cold,
Their passion warms, their eyes are bright.

Oh happy, happy lovers these,
They see the beauty of the night.
I only hear the rustling leaves,
See shadows cast by pale moonlight.

Whilst I upon my couch retire,
And listen to the tick of clocks,
The swains are warm in young love's fire.
They and their damsels chill do mock.

A Wet Summer

Why have the clouds all gathered thus?
Perhaps a plot to stay with us.
Do they conspire to hide the sun
And plunge us into near monsoon.

The whispering surf is hoarse, winds harsh.
Sand castles on the beach are sparse.
Sea mews are tossed across the sky.
They seem to find it hard to fly.

Where Phoebus, is your pleasant smile?
'Tis raining, raining all the while,
And summer's east wind, why so shy?
We long for warmth and azure sky.

Autumn Sadness

Where have you gone, oh summer fair?
Who, pray, enjoys your company now?
Phoebus, a fleeting visit pays.
A frowning west wind spoils his glow.

It scurries clouds across his face.
He peeps at us through windows small.
The once sweet floral woodland ways
Are buried 'neath the leaves that fall.

The poets tell me that you hide
Our cuckoo bird in azure lands
And, also, that his friends abide
With you on balmy, sunny strands.

Oh, hasten back with warm, glad days,
Of picnics, walks and other joys,
Seaside trips to favourite bays,
Long holidays for girls and boys.

After the Storm

The storm has tired, now it sleeps,
Soft winds now smooth, where once gales
Stirred an angry deep.

Too tired now to race and roar,
A sea becalmed brings comfort to the shore,
And walks its journey to and fro.

No heavy-handed waves now beat and bruise the beach.
The gently lapping tides each pebble kiss.
A friendly breeze so softly soothes my face.

White Horses

White horses chase the flowing tide.
Oft times on these, I've wished to ride,
While overhead the sea mews glide.

Whence do these creatures come and go?
We watch the ocean's ebb and flow.
'Tis doubtful if we'll ever know.

Entranced on shore, we stand and stare.
The herd retreats beyond the bar.
Their muted hooves elude the ear.

Where Time Stands Still

I know a place where bluebells grow,
Where time stands still since long ago,
Whilst elsewhere constant changes flow.

Sweet May boughs provide a shade,
A pretty bonnet for this glade,
In autumn, berries for the birds.

Far from the roads, on forest's edge,
By rush-fringed pool among the sedge,
The moorhen builds herself a ledge.

The home of cowslips, daisies pied,
This is where shy violets hide,
Far from the restless human tide.

No new thoughts intrude or spoil,
The rabbits burrow in the soil,
In darkness sings the nightingale.

Wrexham Steeple

Overlooking quiet Temple Row,
A place for the saunterer,
I have watched it and the town grow.

The town can see the time on my four clocks.
It never is in doubt about the wind.
I wear four weather vanes with gilded cocks.

Wrexham sets its clocks and watches by my chimes.
Events and special days, my pealing bells announce.
I recall when Sabbaths were obeyed strictly, were leisure times.

I have a rival here with pleasant sounds.
On Sundays when the people speak to God, and choir sings,
The organist with energetic feet and fingers pounds.

The town was pretty when the stream through Brook St flowed.
Why did they hide the stream, and where?
Some say it runs beneath the road.

From where I sit in lofty chair, I view a constant change,
Of people, buildings, shops and streets.
Some once familiar sights have gone, new ones are strange.

I miss the friendly clatter of the tram,
Which from its Johnstown home, each day,
Brought people to the town, took other people home.

I cannot tell you all in one short chat.
My origin and history are quite a tale.
Come back again another day, to talk of this and that.

The River Alyn

Where tree tops meet, the bluebell grows.
Knee deep in clover, cattle graze.
By ancient Mold, the Alyn flows.

In boyhood day, I sat and gazed.
As Alyn flowed through Loggerheads,
O'er pebbled path where minnows played.

I watched it hasten through the Leet,
By rocky tracks and tranquil walks,
Until it vanished 'neath my feet.

The stream now ran where no man walks,
But suddenly it reappeared,
Out of the ground where darkness lurks.

Moel Fammau with its cloudy hat,
Views Alyn's winding tree-lined course,
And bids it make its way to Holt.

It hurries on to join the Dee,
First having looked at Rossett Mill,
Together, now, they flow to sea.

The River Dee

From Bala lake, white water flows.
Aboil, it tumbles over rocks,
And teases men in their canoes.

The infant Dee is growing now.
At times it saunters leisurely,
Though Corwen sees a hurried flow.

Through quiet acres where flocks eat,
In sunlit meadows by twilight woods,
A place of enchantment, of sensation sweet.

Crow Castle, heathered Horseshoe Pass,
Gaze at the winding River Dee.
Stark, lonely mountains view its course.

Haunt of the shepherd and wanderer,
The Berwyn mountains' peaty heights,
Smile at the Dee, the meanderer.

Llangollen's vale, a pleasant walk,
By Valle Crusis Abbey,
Where nations meet to sing and talk.

An azure bow bends o'er the land.
The river, now, by Bangor crawls,
Casting a whisper on the wind.

The Roman soldier's ghost still rides,
Where once Dee slaked his thirst,
Through ancient Chester, on to meet the salty tides.

Wonderful London

London, gem studded, you walk with grace,
The fount of all things good,
The smile on England's face.

London, the birth place of magic,
You and your wrinkled Thames,
Have witnessed the glorious and tragic.

London, of never-ending fascination,
Your Parliament beating a steady pulse,
The heart of a fortunate nation.

London, with a very English dignity,
Dressed in achievement, history, romance,
Cloaked in a strange, elusive beauty.

London, city of palaces, monuments and treasure,
Home of monarchs and statesmen,
An adventure in knowledge and pleasure.

London, operator of great institutions,
The hub of a busy world,
The envy of many nations.

Tomorrow's World

A privilege God accorded me,
As by a school I chanced to be,
And all the future I could see.

Tomorrow's world unfolded lay,
As children frolicked out to play
And, scores years hence, I viewed this day.

The children all had plans prepared,
I heard them telling Laura Read,
That she must plan, not scratch her head.

'When I retire,' Jimmy said,
'I'll move to Rhyl or Birkenhead,
And fishing go with Uncle Fred.'

'You can't retire, yet, Jim Cole,
First finish school, and then the dole,
Or go down pit to dig for coal.'

A teacher in the door appeared.
A whistle blew, the yard was cleared.
My window on their world was closed.

Beauty

My Friend

My friend, a perfect beauty is,
Not easy to describe in rhyme.
I am in every way bewitched,
By this fair lady's charm.

Her beauteous eyes, in classic face,
Enfolded by her raven locks,
Are but a little peep at this,
The creature of the funny tricks.

She teases me, blows hot, blows cold
I sometimes please, I sometimes vex,
Regards me as a seven year old,
And I am nearly ten times this.

She calls on me to join in play,
Asks my wife this question,
'Can Mr Bob come out today
To play "I Spy" or skipping?'

She gives me lessons in her school.
To teacher I must listen.
'Behave yourself, Attention please,
Or you'll stay in at playtime.'

'Pick that up,' 'Go there,' 'Come here.'
She keeps me very busy,
Then suddenly decides to leave.
'I'm going to see my mummy.'

Which, Leila, Sara, Jessica or Kiera?

Days of sweetness shall be yours,
If my wish be granted,
Honeyed pleasures dress the hours,
Your path be brightly painted.

Thus structured days will match your grace,
Their music frame your beauty.
The matchless rose reflects your face,
Sweet as angelic infancy.

The syrups of the world are stored,
In twin pits called your eyes.
My thoughts of you are pure gold.
They float on life's sweet, secret sea.

No shadow is cast, or reflection by you.
How could these appear to mortal sight,
For you are a star,
And I am in love with the night.

Of Julia (Robert Herrick, 1797–1856)

I quote, I quote Bob Herrick, who,
Wrote lovely words of Julia.
I wonder if, had he seen you,
Would he have penned of Julia.

This silk-clad Julia, so they say,
Was a lass beyond compare.
Her every aspect pleased the eyes,
Of all who saw her beauty rare.

How sad that Herrick did not paint,
That he lacked skill with oils and brush,
Did not reveal this Julia saint,
She must for ever be hush-hush.

Beauty

On the forest's edge, I glimpse a handsome jay,
His coat multi-colour, imitating Joseph,
Beauty in plumage, outstanding in a world so gay.

Whither, who knows, with his mocking cackle, he disappears.
Sweet singing artistes take the stage.
Bright coloured flocks of finches dine on thistle ears.

The blind man wonders, when you tell of these.
Be glad, that blest are you with sight.
You have seen the pretty flowers, birds and trees.

The Tree

Oh, monarch of the woodland ways,
Your green-clad boughs the traveller shades,
From passing showers or scorching rays.

With outstretched arms, you welcome all,
Who dwell within your leafy wall.
Inside your bark, the insects crawl.

Your ivy wraps, the small wren hides.
Likewise, the robin's brood abides.
The wise old owl in you confides.

The busy rooks adorn your crest.
Woodpeckers tunnel to their nest.
A cuckoo stays a while to rest.

Beneath your roots, the badger's home,
Of passages and sleeping rooms,
A palace hidden 'neath your throne.

Cherry Trees

Sweet views of spring are with us now.
The cherry trees wear feathered bough.
Farewell to winter's ice and snow.

Anticipating future feasts, the blackbirds sing,
In airs made warm by sunny spells,
As Phoebus smiles, encouraging the spring.

The whispering spruce and poplar tall,
Are ornaments that please all eyes.
The apple tree boasts mistletoe for Yuletide ball.

With hidden birds' nests in her hair,
The ivy's arms embrace the orchard walls.
Do these with cherry trees compare?

A cherry cordon lines the orchard wall.
With glee the blackbird pipes, 'There is no net.'
Of cherry ripe, the gardener will not gather all.

Perhaps this gardener, as did Addison of old,
Prefers the blackbird's song to cherry pie,
And, for a song, he pays in fruity gold.

The Mountain

The raven clasps your rugged walls.
The stag seeks shelter in your halls.
The wild goats leap your waterfalls.

The eagle soars above your crown.
The bees inspect your purple gown.
The curlew calls in tongue unknown.

The winds lilt o'er your shoulders high.
The storm clouds hide you from my eye,
The signal that the rain is nigh.

The shepherd climbs to tend his sheep.
The playful lambs do run and skip.
The dog a watchful eye doth keep.

The Lords

Their Lordship's House

The mistress told another tale,
Of happenings at the bishop's place.
His housekeeper, Louisa Small,
Related this to Nellie Wallace.

This Nellie is Peek's daily maid,
A well-informed Miss Wallace.
Apart from being a priceless aid,
She tells us tales of Horace.

Bishop Horace in the Lords,
Produced a sturdy watch.
Its tick annoyed the sleepy bards.
They began to moan and twitch.

Lord Stockton made a speech on clocks.
He touched on watches, too,
The bishop gave him quite a shock.
He began to hiss and boo.

The Lords were piqued, they warned him,
Withdraw, or else get out.
The Bishop promptly told them,
'I'm off, it's party night.'

Parties at the bishop's are always jolly fun,
Sweet music, laughter, shuffling cards,
Food in plenty, flowing wine,
Far better than the House of Lords.

Back at the Lords

The bishop having made his peace,
With Stockton and the other Lords,
Resumed his seat in quiet bliss,
Determined hence to watch his words.

There was no need for him to speak,
To irritate the Lords.
His monster watch would do the trick,
And wake each sleeping bard.

The watch we mentioned was the cause,
Of Stockton's speech on ticking things.
This irked old Horace, much, because
He growled and muttered words with stings.

This disturbing watch brought forth black looks,
Drowned his acid words with ticks,
Buried protest 'neath the tocks.
'Twas like a factory making tacks.

The bishop, having had his fun,
He liked to level scores.
He settled down despite the din,
And slept with noisy snores.

Guy Fawkes

Upset one budget day, Guy Fawkes
Had left the House in angry mood,
And homeward bound decided, to use his fireworks.

Those hooligans, he promised, a lesson he would teach,
Because of their loud bleats of 'Shame',
Guy had not heard the speech.

The Heffer types and Wedgie Benns,
Who stamped their Co-op boots,
Were obstacles to progress, who never spoke good sense.

So Guy Fawkes set his plans afoot,
Some pyrotechnic noise and smoke,
To cure suede or leather shoes, and also Co-op boots.

The peaceful realm, not ruled by fools,
That Guy Fawkes had in mind was doomed,
For interfering policemen took umbrage, and Guy's tools.

Eluding lawmen, wily Guy, now sped to Huddersfield,
Where he set up Standard Fireworks,
Who made bangers for the world.

To lands of note, exported to lesser places, too,
Guy's efforts cured governments,
From China to Peru.

Some claim that spacemen were first put up by Guy,
Though sudden bangs made many jump,
'Tis doubtful if they reached the sky.

Simple Pleasures

The Simple Life

Those who common pleasure prove,
The simple life do lead.
Their wine, the showers from above,
The rivulets provide their mead.

Unlike the high born cavalier,
Astride his mount in fine array,
They taste the pleasures of the year.
On sturdy limbs they walk the day.

They seek not honour, fame, or wealth,
But humbly toil they for their bread,
Rejoicing in their ruddy health,
Aware that they are blessed by God.

Oft ill-shod and scant of vest,
Caring little for their purse,
Happiness is theirs who taste,
Good company and well-written verse.

You Have Not Seen

You never walk across the heath,
The windswept doorway to the moor.
Where I climb stiles, you walk beneath
Tall chimney stacks that skyward soar.

You window-shopped your way from school.
I slowly dawdled by the lake,
Made stony stitches on the pool,
Saw ducklings in their mother's wake.

You have not seen the far off hills,
Reclining in the summer haze.
Your views are stunted by the mills,
And sunlight plumes that spoil your days.

You have not seen the curlew's beak.
He never visits you in town.
He haunts the moorland wild and bleak,
Or mountain clad in purple gown.

Come dwell with me, fair city maid,
Where bends heaven's curtain, azure blue,
Where chestnuts tumble on your head,
And every day brings something new.

Come, let me show you prettiness,
My tree house hidden in the woods,
The lark's nest hidden in the grass,
And toys to please your varied moods.

The Pinks

From Ralph Tuggie's garden
In flawless beauty cast,
Came pinks to make the garlands,
To decorate the feast.

Oh, flower perfect, 'Sops in Wine',
Some country folk have named her.
To ancient Greeks, 'The flower divine',
Carnations call her sister.

In Tudor times Ralph Tuggie grew,
The finest gillyflowers.
The many blooms of varied hue,
Spiced the July hours.

As lovely as these gillys were,
They did in no way match the pinks,
Whose spiced perfume filled the air,
And gave the bees the sweetest drinks.

Bacchus

To Bacchus, thanks for merriment,
The golden cup, the cup that cheers.
Let wine dispel your heaviness,
Sip now the grapes fermented tears.

The toast is 'Bacchus', God of wine,
But let us not forget milk stout.
Whilst all applaud that splendid vine,
Let drinkers all give hops a shout.

The nectar of the glens, by some
Is held in very high esteem,
And they who drink diluted rum,
Would not be happy with poteen.

The Japanese make wine from rice.
The Danes and Germans lager sip.
Some drinkers spoil good gin with ice,
But all, to Bacchus, glasses tip.

Not all agree that scotch is wine,
Being whisky and distilled from grain.
Historians tell of Gods and wine,
But quite a lot they don't explain.

The poets tell me that the goblet's dew,
Did Bacchus first ordain.
Did he the hop fields' harvest brew,
And quench the thirsty lands with rain?

May Day

'Tis May Day, merry, full of fun,
A cloudless sky and glorious sun.
Today, we celebrate the spring.
The villages with laughter ring.

Each village green will, today, see
The lads and lasses full of glee,
Dancing to the fiddler's call,
Around the Maypole, one and all.

The children each will have a treat,
At tables laden for the feast.
All mums and dads will busy be,
Serving food and pouring tea.

Cool nettle beer for morris men
And others who have races run,
The vicar will award a prize,
To winners with young eager eyes.

The granddads yon great oak,
Will criticise and puff out smoke.
'These 'ere young 'uns aren't as fast,
As we were in times gone past.'